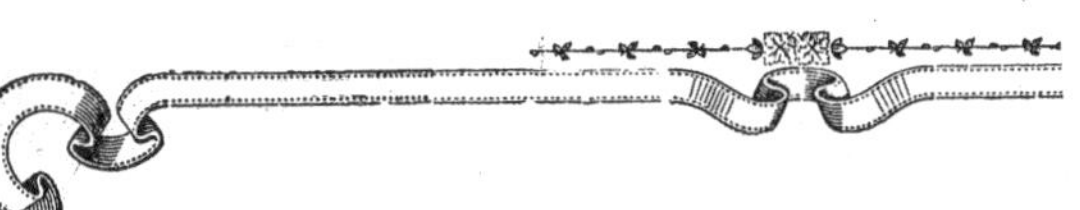

STEAM ON THE CANALS.

REPORT

MADE TO

HON. PETER COOPER

BY

HUGH McKAY,

Superintendent.

NEW YORK:

EVENING POST STEAM PRESSES, 41 NASSAU STREET, CORNER LIBERTY.

1875.

STEAM ON THE CANALS.

BROOKLYN, NOV. 28th, 1874.

HON. PETER COOPER:

Sir,—In preparing a report of the work done by the canal-steamer *Peter Cooper* and consort *Edith*, during the present season, it will be just and proper to compare their performance with other boats running in the same trade, not invidiously, but critically. To prove that the *Peter Cooper* system is the best, not only to double the capacity of the canals, but the one that can be introduced with the least expense to the present boat owners, as every other first-class boat now on the canal can be made a part of the system without any alteration or expense. This was made an important requirement of the prize law of 1871, and a majority of the competitors for the prize were fully impressed with the idea that it was an indispensable condition of law that the present boats should be used. To satisfy this requirement the *Peter Cooper*, a boat that had run part of the season of 1873 as a horse-boat and part as a stern-wheel steamer, was again altered upon a plan which I submitted to you in January, 1874, the success of which, upon the Erie canal this last season has demonstrated that any ordinary marine engine and screw propeller, placed in a proper positionf or work, in a narrow and shallow channel, with only a slight alteration of the stern of the present boats, could comply fully with all the requirements of the prize law and would be a commercial success providing she had an engine of sufficient power to tow another boat. In this lies the success of the *Peter Cooper* system, and the only reason why steam-towage has not heretofore proved a success, is because the proprietors of the different methods of towage have tried to tow too many boats, not taking into calculation the time lost in locking the extra boat. It has been demonstrated by the *Peter Cooper* the past season that, on her first trip down, with a boat in tow, 54 of the 72 locks were ready for both boats; the time of locking both boats 7½ minutes. On her second trip 60 locks were ready for both boats. On her third and last trip 40 locks were ready for both boats.

In going up we found it saved time, owing to the leakage in locks, to lock one boat after the other in the same lock. Time of locking both boats 12 minutes, so that the delay at the locks in towing one boat is practically nothing.

The time employed in passing through the 5 combined locks at Lockport was 25 minutes. The *Peter Cooper*, as you are aware, is an ordinary full-built canal-boat; her alterations consisting in sloping up her stern from a point 20 feet forward of her stern-post up to her 6-foot water-mark, so as to give her easy steering capacity and solid water to her wheel. The only peculiarity in the construction of her motive power is the inclination of her shaft, which slopes down from the engine, forming an angle of 12 degrees, with her keel thus lessening the slip of the propeller materially, as the current of water flowing from the wheel strikes against the bottom of the canal and backs up under her stern, thus preventing her wheel from drawing the water from under the bottom of the boat, a serious trouble in shallow waters when the shaft is placed parallel with the keel. The *Peter Cooper* is also so easy steered and handled that either night or day we found no difficulty in entering locks or avoiding collisions, and on any of the levels where the water was of the regulation depth, we found no difficulty of attaining a speed of 3 miles per hour, with the boat in tow. On our last trip down we overhauled and passed 84 boats that had left Buffalo one and two days before us. The method of passing horse-boats adopted on the *Peter Cooper* is, I think, worthy of mention, as one of the principal objections the horse-boatmen have to steamers is that they have to stop to let the steamers pass. This in itself would not cause any great inconvenience; but as the bow of the horse-boat catches on the bank the suction of the passing steamer draws the stern around and across the canal, often catching it on the opposite bank, thus completely blocking the canal for the time and apt to be attended with serious consequences, for if this should happen on a short level, where the swells rapidly pass to and fro, the boat would be seriously damaged by hogging. But even where no damage occurs it tries the patience of the boatmen and more or less injures their teams by the extra strain put upon them on starting. This is all the fault of the steamer, and it is no wonder that animosities are created. The *Peter Cooper* on the other hand has proved a benefit and makes friends instead of enemies by having a strong line ready to hitch on to the bow of the horse-boat as she catches on the bank with which we jerk her off, giving her the same speed for the moment as ourselves, and straightening her along the canal, thus keeping her out of the way of the boat astern and giving her headway enough to take all the strain off her team. That this is appreciated by the boatmen is shown by their willingness to lay by the second time we meet them. By helping others we help ourselves, as we are not delayed a moment, not even finding it necessary to slow-down the engine, for the horse-boat's bow catching on the bank has the effect to throw the wave of replacement between the boats, forcing them apart instead of sucking them together. There is no difficulty at all with the boat in tow, she getting past without any inconvenience; but where two or three boats are strung in a line four or five hundred feet astern the case is very different; then, difficulties or collisions are liable to occur.

Capt. Henry Spur, of the steam canal-boat *Henry L. Fish*, found so much annoyance and expense in towing three boats that he abandoned it before the season was over. Capt. Petre, of the *Eureka*, towing two boats, finds, by experience, that he has one too many. The *Cooper*, however, has been a success, because there being sister-locks the whole length of the Erie Canal, it is evident that when both locks are ready the steamer and her consort can pass through in the same time, the steamer, however, coming out of her lock first, for the purpose of gettiug into position.

If there were three locks abreast, as far as locking was concerned it would take no longer to lock three than it would one. There being but two locks, the train is limited to two boats.

The advocates of the towing system, although the most practicable men on the canal, have heretofore failed to bring their ideas down to towing one boat; every man that has had anything to do with a steamboat knows that it will not pay to carry but 200 tons of cargo 500 miles in a steamer at the present rates. A steamer alone is only profitable on short canals, where boats are loaded and unloaded ten times where an Erie canal boat is once—it is, therefore, absolutely necessary to increase the load without materially increasing the expense; this, after all, is the only problem for a steam canaler. The problem was said to be solved by the advent of the *William Baxter*—but the *Baxter* plan, as exemplified in the *William Baxter*, is as defunct as her competitors, the late *Montana* and *Port Byron*. It was claimed for the *Baxter* that her consumption of fuel was less than oats for a mule, but on looking over the reports I find it variously stated at 14, 18 and 31 pounds per mile, which figures are right I cannot say, so what she really did burn per mile, as far as the public is concerned, may be considered an unknown quantity; what she cost for repairs is rather more definite, namely, $2,481.70, yet there are two items in the bills that are rather cloudy (see last report of steam on the canal by Chief Engineer D. M. Greene). The items are—cost of two propellers is charged as only $46, yet the cost of putting them on is charged as $750. So with all this uncertainty about her expenses, coupled with the fact that her inventor has abandoned the crude idea with which he entered upon the problem, it would not be fair to take her as a comparison with the *Peter Cooper*, as Mr. Baxter has so altered his later boats that they in fact are a new departure entirely, with the exception of the *Model*, which is still similar to the log bilge boats now used upon the New Jersey canals. It will be fair, therefore, to compare the performance of the *City of Rochester*, since her advent, with the work done by the *Peter Cooper* and her consort during the same time, estimating the running expenses of both steamers the same, with the exception of the fuel consumed, and adding to the *Peter Cooper's* expenses the cost of running a boat like the *Edith*. The delays incident to canal navigation I have not taken into account. I think it just, however, to state that the *City of Rochester* had preference at the locks, which gave her a decided advantage in time, which is decidedly wrong. I do not think that steamboats, with or without tows, ought to have any preference on the canal; if steamboats cannot run and succeed under the same rules that govern other boats, they have no business on the canal. I think it is also proper that the time occupied in New York by the *Peter Cooper* on demurrage, and the time occupied in Buffalo and on the dry dock repairing hull and painting the *Peter Cooper*, should be substracted from the total time, as the boat was under no expense for a crew.

The following is a correct record of the number of trips made by the *Peter Cooper*, from the 6th day of August to the 22d day of November, 1874, together with her expenses and number of tons of freight carried during the same time, to wit:

Total number of days	109
Substract days on demurrage and Dry Dock	26
Total days	83
Number of trips	3
Total number of tons carried by official clearances	1,818

EXPENSES.

1 Captain, at $2 00 a day for 83 days	$166 00
2 Engineers, each 2 00 " " 83 "	332 00
2 Steersmen, " 1 17 " " 83 "	194 42
1 Bowsman, " 0 83 " " 83 "	68 89
1 Cook, " 0 50 " " 83 "	41 50
Provisions for 7 persons, including fire and light, $4.55 a day for 83 days	377 65
Oil and waste, $15 per trip, 3 trips	45 00
Coal used—100 tons—average $6	600 00
Expenses of boat *Edith*—crew and maintenance—$140 per month, for 83 days	387 41
Interest on cost of boat, $4,000, at 7 per cent	70 00
Total	$2,282 87

DEDUCTIONS.

Number of tons carried in 3 trips	1818	
At a cost		$2,282 87
Or about $1.25 per ton.		
Average number of days to a round trip		$27\frac{2}{3}$
Tons of freight carried West, 2 trips		501
Tons " " East 3 trips		1,317
Number of bushels wheat carried East		43,838
Number " " by *Baxter* boat		27,207
Difference in favor of the *P. Cooper*		16,631

The following is a record of the Baxter boat *City of Rochester*, from the 20th day of August to the 20th day of November, 1874; total number of days, 92; number of trips, 4; average number of days per trip, 23; total amount of freight carried by official clearances, 1,163 tons.

ESTIMATE OF EXPENSES.

1 Captain, ... $2 00 per day for 92 days	....	$184 00
2 Engineers .... 2 00 " " 92 "	....	368 00
2 Steersmen ...... 1 17 " " 92 "	....	215 28
1 Bowsman .... 0 83 " " 92 "	....	76 36
1 Cook .... 0 50 " " 92 "	....	46 00
Provisions, including fire and light per man—65 cts., for 92 days	.....	518 60
Oil and waste per trip, $15×4	....	60 00
Coal per trip, 20 tons 20×6=120, 120×4=	....	480 00
Total	....	$1,948 24

DEDUCTIONS.

Number of tons carried in 4 trips	..1,163	
At a cost of	....	$1,948 24
Or about $1.67 per ton.		

REMARKS.

We find therefore, that it cost by the *Baxter* system to carry 1 ton of freight between Buffalo and New York	$1 67
By the *Peter Cooper* system	1 25
Difference in favor of the *Peter Cooper*	$0 42

Compared with horse-boats as per report of D. M. Greene, the deductions are as follows:

Baxter boat, per ton	$1 67
Horse boats, per ton	2 02
Baxter boat cheaper than horses	$0 35
Peter Cooper system, per ton	$1 25
Horse-boats, per ton	2 02
Peter Cooper system cheaper than horses	$0 77

With the record above given, it will be seen that the *Peter Cooper* has succeeded in meeting all the requirements of the prize law.

1st.—That the invention or device shall be tested and tried

at their own proper cost and charges of the parties offering the same for trial.

2d.—That the boat shall, in addition to the weight of the machinery and fuel reasonably necessary for the propulsion of said boat, be enabled to transport, and shall actually transport, on the Erie Canal, on a test or trial exhibition, under the rules and regulations now governing the boats navigating the canals, at least 200 tons of cargo.

3d.—That the rate of speed made by said boat shall not be less than an average of 3 miles per hour, without injury to the canals or their structures.

4th.—That the boat can be easily stopped and backed by the use and power of its own machinery.

5th.—That the simplicity, economy and steerability of the invention or device, must be elements of its worth and usefulness (this and the following requirement are especially answered by the *Peter Cooper*).

6th.—That the invention, device or improvement can be readily adapted to the present boats.

7th.—That the commission shall be fully satisfied that the invention or device will lessen the cost of canal transportation.

8th—That the general adoption of the invention or device would increase the capacity of the canal.

9th.—That the competing boats shall make three round trips between Buffalo or Oswego and New York.

Having answered all the requirements of the prize law, we therefore claim for the *Peter Cooper* the $100,000, as being the only boat that has done so.

The system adopted in the *Peter Cooper*, although not requiring any very eminent original engineering ability, it necessarily cost a large sum of money to demonstrate its practicability, and as it is the only system that the practical canalmen look with favor on, it should be suitably rewarded, inasmuch as it is made free to all, and not restricted in its use by being a patent in the hands of a monopolizing company, who grant royalties to only them who can afford to pay them for the right.

The *Peter Cooper* was altered and her machinery inproved for the season of 1874, because it was believed that the powers of the commission appointed by the prize law did not expire till the latter part of the season of 1874. The Legislature of 1873 amended the original law so as to continue it in force until that time, and believing that the *Peter Cooper* could win the prize, we ran until she had fulfilled the law.

The *Peter Cooper* was run till the close of this season, carrying on her three last trips almost double the cargo east and west than any other steamer did in the same time, and more than double the quantity required by the law, and the *Peter Cooper*, with her consort, is ready to run a whole season, under official observation, against any single steamer of any other system, to prove her championship, barring all accidents and delays of any kind whatever, it being supposed that the same delays are incident to all.

I think it proper in this report to call your attention to the enormous expenses in the shape of tolls, commissions, insurance, elevations and storage that the internal commerce of the State of New York is burdened.

1st.—For the privilege of doing six months' work on the canals, a first-class boat must pay:

To the State, for tolls	$1,764 00
To insurance companies	245 00
To commission merchants or scalpers	175 00
Trimming, tugging, &c.	175 00
Elevation	350 00
Shortage	210 00
Total	$2,919 00

Nearly $3,000 above the actual running expenses of the boat.

Of the above tax, the most onerous item is the shortage or

the difference in weight of a cargo of grain passing from one elevator through another.

The scales may be wrong, the hoppers may leak or the grain stick in some mysterious manner to the elevator, yet the boatman has no redress. He must either agree to pay for the shortage, or remain without a cargo; and what is very curious about this transfer of grain is that there is never any longage.

A case happened but a short time ago where a load of grain from Chicago unloaded into four canal-boats at Buffalo, for New York. The schooner fell short of her load nearly 100 bushels. The boats receipted to the elevator at Buffalo for their load of course, less the hundred bushels, yet each boat fell short on being elevated out at New York—two boats 20 bushels each, two boats 34 bushels each, making a total of 208 bushels on one cargo loss in passing through two elevators. In this manner there is about 60,000 bushels of grain paid for by the canal-boat owners every season. This is all wrong. A boat owner or captain should not be made to pay for shortage on a cargo that he cannot check into or out of his vessel, if he is willing to make affidavit that he delivered all the cargo he took in. This should be made a law of the State.

Tollage.

That the tolls on grain are too high is now admitted by everybody interested either directly or indirectly in the prosperity of the State, and the enterprising merchants who eighteen years ago loaded the schooner *Dean Richmond* with a cargo of 14,000 bushels of wheat, at Chicago, direct for Liverpool, are only waiting for the Canadian canals to be enlarged to repeat the experiment with larger cargoes, and thus flank New York completely.

As a comparison of the cost to the *Dean Richmond* of 14,000 bushels of wheat passing through the Canadian canals to the ocean, with the cost of the same amount through the Erie canal, from Buffalo to New York, will show better than anything else that our tolls are excessive. I will give it.

From Port Colborne, the entrance of the Welland Canal, opposite Buffalo, to the ocean at the Straits of Belle Isle, the distance is about 1,351 miles; the charges are as follows, on vessel and cargo :

Towage on Welland Canal	$35 00
" between St. Lawrence and canal	50 00
" through three canals	24 00
" from Montreal to Quebec	60 00
Pilotage to sea	30 00
Total tolls on vessel and cargo	160 00
Total	$359 00

The same amount of wheat shipped from Buffalo to New York, for State tolls alone, would call for an outlay of $420, or $61 more than the total charges of vessel and cargo by the St. Lawrence route, and yet be 1,000 miles further from its destination, if bound for Liverpool, than at the mouth of the St. Lawrence.

Is it any wonder, then, that the merchants of the West are seeking other and cheaper routes?

Tolls should be collected on the boat, at so much per mile, and let her carry more or less, disregarding the quality or quantity of her cargo. This would abolish weight locks and all the expenses of collection and officers. The captain could be furnished with tickets to be punched at every lock, which would be correct evidence of the number of miles sailed. A simple system of this kind would save thousands annually.

The terminal facilities of New York, with a very little outlay, could be made the best and cheapest in the world, as every boat that comes in from the West is a floating warehouse, with free storage for three days, that can be taken alongside of the steamer at her own pier, so that while the steamer is discharging the foreign goods on one side, the grain to pay for them, is pouring in at the other from the floating elevator; so that by the time the steamer is unloaded, she is ready for sea, with her return cargo on board. This great facility would be increased a hundred-fold by building a pier one or two thousand feet in length and fifty feet wide, in the Hudson river, midway between the New York and New Jersey shores, for the canal-boats to moor to—loaded boats on one side waiting for orders, and light boats on the other, making ready to tow up the river. A pier of this kind, built on iron cylinders, filled with concrete, would be no obstruction to the flow of the tides, and would be a great blessing to the port in winter, by splitting into pieces the immense fields of ice that block the harbor during the frosty season. Transfer elevators could also be built upon it capable of handling millions of bushels of grain with the utmost dispatch.

With the hope that these concluding remarks will not be considered out of place in addressing a prominent New York merchant,

I respectfully submit this report,
HUGH McKAY.

APPENDIX.

As this report was being printed the message of Gov. S. J. Tilden to the Legislature appeared in the papers, and while it foreshadows a truly intelligent, liberal and comprehensive canal policy, I wish to call your attention to an error that may be fatal, not only to the Erie Canal, but also to the commercial prosperity of the whole State.

The Governor, in speaking of tolls, says : " The present tolls on wheat are 3 1-10 cents, and on corn 3 cents per bushel, from Buffalo to Troy, 345 miles ; they were reduced in 1870, those on wheat, from 6 21-100, or one half, and those on corn, from 4 83-100 to 3 cents, or about 38 per cent.

One cent per bushel taken off the present tolls, and the same proportion on other articles, would annihilate nearly all the net income of the Erie Canal considered alone, and would make a deficiency in respect to the four canals retained of $500,000 a year, if the expenditure should be the same as the three past years."

I do not think that canalmen advocate the reduction of the tolls on anything but grain, which are well known to be too high, so high as to gradually force the grain into other waters ; the grain business of the canals has remained nearly stationary for about three years, while that of the Trunk Railroads have increased enormously. The New York Central, in 1871, landed in New York, in round numbers, 3,000,000 bushels; in 1874, 16,000,000 altogether by the Erie, the Pennsylvania and New York Central; the grain landed at New York was, in 1871, 19,000,000; in 1874, 40,000,000. During the season of 1874, fully one-half of the boats owned on the Erie Canal were tied up; there are 2,345 boats on the underwriters' books that will insure for grain. Admitting that there has been shipped through the canal this season 56,000,000 bushels of grain, which is over the average, it would *not give three loads to each boat*, while she is capable of carrying seven during the season; this shows that the Erie Canal is too large, or that it does not get its proportion of business. Canalmen think that if the tolls are reduced, that the receipts for tolls will be about the same, on account of the increase of business that will naturally seek the least expensive route, unless the tolls on grain are reduced. From the experience of the past season, I fear, that the grand Erie will be *ticketed to let.*

I do not see what better proof is wanted that something is wrong than the fact that the capacity of the present grain boats are able to transport, during the season, 131,320,000 bushels, but yet do not transport more than 56,000,000.

It is difficult to see any virtue in a policy that makes a boat-load of wheat pay $240 to pass through the canal, while the same weight of coal only pays $70; surely the wear, tear and cost to the canal is the same. I hope that you will use your influence in favor of reduced tolls; it is not merely a question of a few dollars, but the entire canal commerce that is at stake.

Yours respectfully,

HUGH McKAY,

Superintendent.

www.ingramcontent.com/pod-product-compliance
Lightning Source LLC
LaVergne TN
LVHW020644110826
845149LV00004B/1347

* 9 7 8 1 4 1 8 1 8 9 8 3 9 *